Millau Viaduct

The Deutsche Bibliothek holds
a record for this publication in the
Deutsche Nationalbibliografie;
detailed bibliographical data can
be found under http://dnb.ddb.de

Library of Congress Control
Number is available

Prestel Verlag, A Member
of Verlagsgruppe Random
House GmbH

Prestel Verlag
Neumarkter Str. 28
81673 Munich
Germany
Tel +49 (0)89 4136-0
Fax +49 (0)89 4136-2335
www.prestel.de

Prestel Publishing
900 Broadway, Suite 603
New York NY 10003
USA
Tel +1 (212) 995-2720
Fax +1 (212) 995-2733

Prestel Publishing Ltd
4 Bloomsbury Place
London
WC1A 2QA
UK
Tel +44 (020) 7323-5004
Fax +44 (020) 7636-8004
www.prestel.com

ISBN 978-3-7913-4687-8

Millau Viaduct Foster + Partners

Norman Foster
Thomas Leslie

PRESTEL
MUNICH · LONDON · NEW YORK

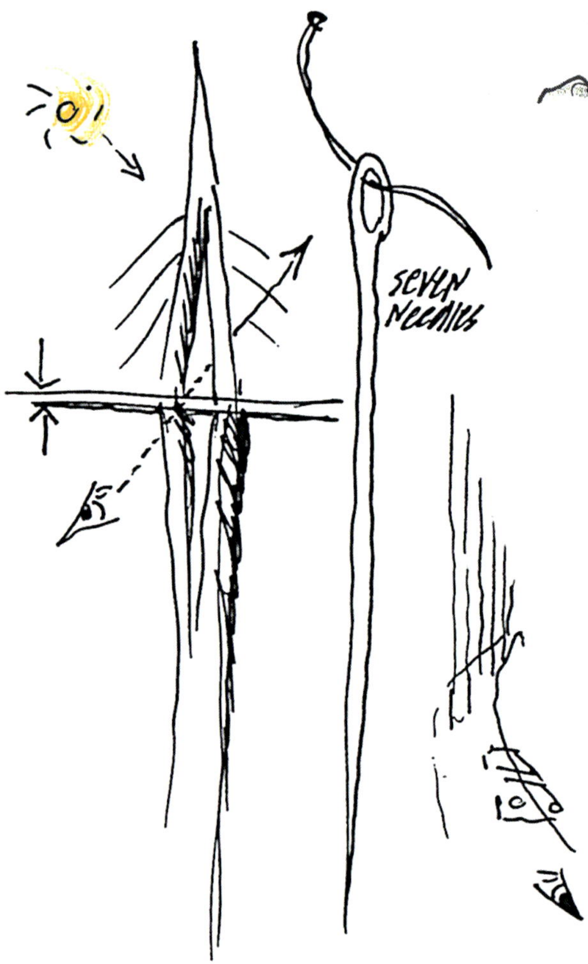

In this series of sketches, Norman Foster explores ways in which the bridge and its constituent parts can be lightened – both physically and visually – and made to recede when seen against the sky.

Overleaf: The bridge's seven masts appear to hover above the cloud line.

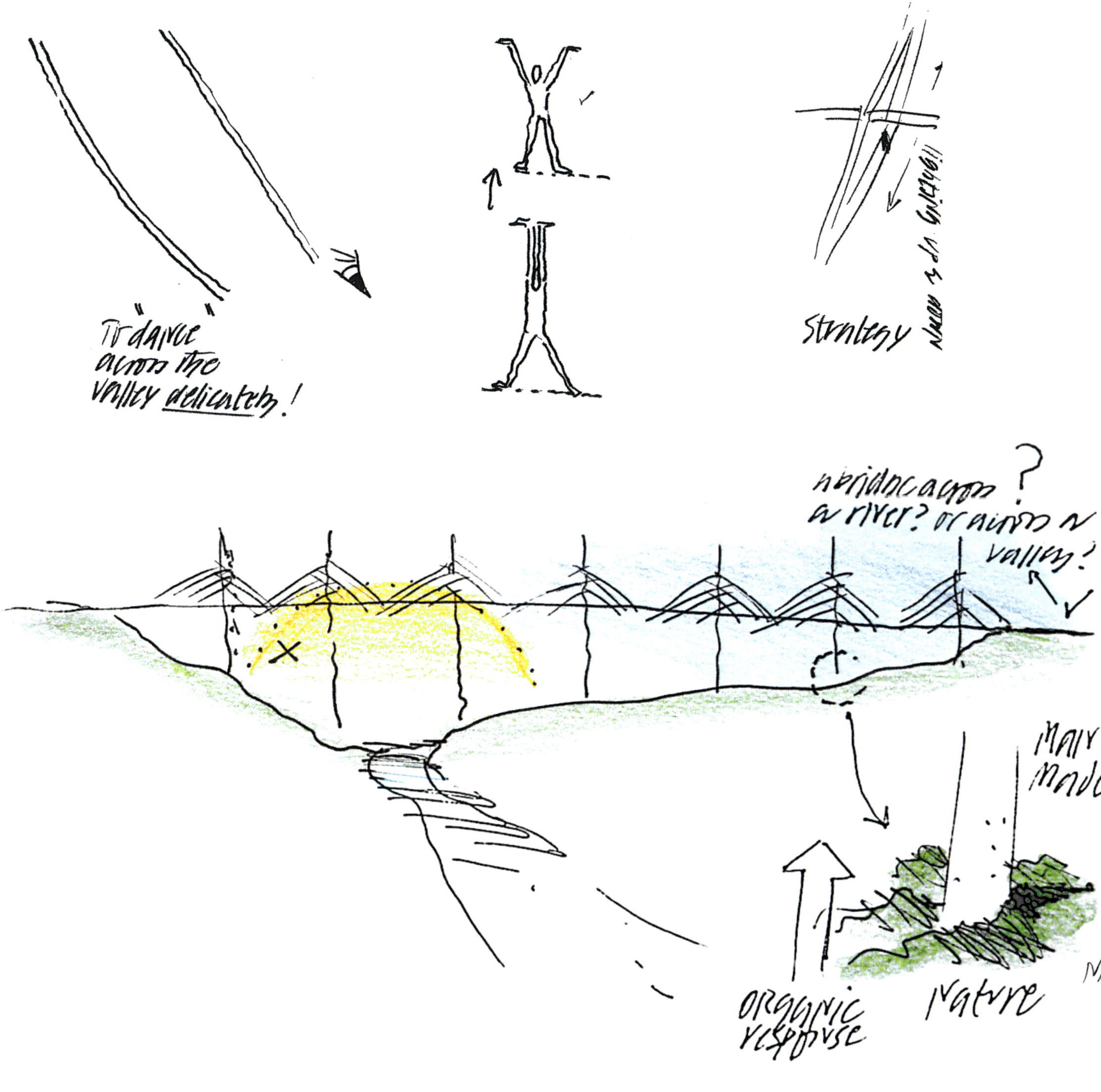
To "dance" across the valley delicately!
strategy
a bridge across? a river? or across a valley?
man made
organic response
nature
NF

Introduction Norman Foster

The design of bridges is usually thought of as belonging more to the realm of the engineer than to that of the architect, but the architecture of infrastructure can have a powerful impact on the environment. Like our Millennium Bridge over the River Thames in London, the Millau Viaduct illustrates how the architect can play an integral role in expressing the relationship between function, technology and aesthetics in a graceful structural form.

The Millau Viaduct was commissioned by the French state to complete a missing link in the A75 autoroute from Clermont-Ferrand to Béziers across the Massif Central in southern France. The competition brief was for 'a bridge across the River Tarn'. However, the Tarn is quite narrow at this point – you almost feel that you could jump across it – but it runs through a spectacular gorge between two high plateaux: the Causse Rouge to the north and the Causse du Larzac to the south.

Based on how one read the topography I saw that there were two ways one could approach the design of the bridge. One could say that there was a case for making a big span over the river itself, because the river was the historic generator of the landscape; and that would lead to a very recognisable engineering solution. Or one could say that it was not about spanning the river at all – that the real question was how to span the 2.46km between the two plateaux in the most elegant and economical manner. The river, in this reading, is only an incident along the way.

Interestingly, most of our competitors based their designs on the first interpretation, with a primary symbolic span over the river and smaller approach spans to either side. The divide between us and them was essentially a philosophical, not a technical one – although our philosophy would eventually deliver real technical advantages.

I felt so strongly about this that my engineering colleagues insisted that I should lead the presentation in Millau. I began by setting out these two approaches. I said that I was definitely not the architect they should choose if they thought they wanted to build a bridge across the river. I explained that we wanted to create a structure that would address the grandeur of the entire valley, a structure that would leap across the landscape from plateau to plateau.

Later, after we had won, I discovered that one of the key differences in the way we presented our proposal was that all the other team presentations were led by the engineers. Firstly, the engineer would say something and then the architect would speak. Ours was the only team where the architect set out an overarching philosophy that encompassed all aspects of the design – something that had greatly impressed the jury.

From the beginning our approach was about collaboration – getting the team around a table and discussing ideas in a creative way. We wanted to explore with Michel Virlogeux and his engineering colleagues what was really possible – to find the architecture within the engineering. I have no doubt that the success of the bridge – structurally and aesthetically – depended on the integrated nature of the design team. Michel in particular was a completely kindred spirit.

There is an intellectual and aesthetic purity to the concept of the gentle arc of a ribbon of road suspended from needle-like pylons planted in the landscape. It is both poetic and ruthlessly logical, lending itself to optimised and repetitive spans and in turn to improved economy and buildability and therefore reduced financial and logistical risk. These were the welcome benefits that flowed from clear strategic thinking.

The idea that you can build over such a great distance – with pylons exceeding the height of the Eiffel Tower – is to me astonishing. Just as a technical feat it is quite extraordinary. I have almost lost count of the records we broke: the viaduct has the tallest piers and the highest bridge tower in the world; it has the highest road deck in Europe; and at the point where it crosses the river the deck is 270 metres above the water, higher than our Commerzbank tower in Frankfurt – at that time still the tallest building in Europe. It so happens that highest part of the road deck and the tallest pylons correspond with the tiny path of the river!

Inevitably, there were many detail challenges because the wind-forces at such a great elevation are incredible; and the piers have to accommodate the enormous accumulated expansion and contraction of the road deck. To achieve this, each pier splits into two thinner, more flexible columns below the roadway, their tapered form both expressing their structural loads and minimising their profile. Above deck level, steel masts correspond with the concrete piers, forming A-frames which are cable-stayed, the cables placed in a fan arrangement between the carriageways.

Although the concept was very clear, as the detailed design progressed many difficult decisions had to be made. For example, originally we thought that the road deck would be triangular in section, but wind tunnel tests showed that this was not ideal at such an altitude. In fact it was deeply problematic. Michel suggested we adopt a trapezoidal deck, which provides negative lift in strong wind conditions – as he had done on the cable-stayed Pont de Normandie. When we studied the options we discovered that the trapezoidal deck not only performed better but looked better too.

Then there was the question of what colour to paint the cables. Should they be dark, so that they virtually disappeared when seen against the green of the valley, making the bridge look as if it was standing under a spell? Or should they be light and so less prominent against the sky? A sculptor will tell you that if you want a piece to recede in the landscape it should be dark. I remember a conversation with Henry Moore who said that a dark bronze colour would do this. To achieve the opposite effect and give the piece prominence it should be light. In the end, because the cables are seen against the sky, I decided that they should be white and therefore recede.

We were guided constantly by a desire to reduce everything to its essentials – structurally and visually. This was in part a response to the quite justifiable fears expressed by some people at the time of the competition that a bridge would spoil the wonderful landscape of the Tarn Valley. But I also wanted the structure to look as delicate as possible, almost floating, so that seeing it you would marvel at how it was possible. I wanted the piers to look as if they had sprouted naturally from the soil, not been imposed upon it. We modelled numerous options for the pier profile, gradually refining its form until we had a very pure shape. Finally we introduced a vertical groove, which emphasises its verticality. We also made extensive modelling studies of all the components of the deck: the cables, the piers, the wind shield, and the junctions between them.

We strove from the beginning to make the experience of driving across the bridge special – I wanted motorists to feel as if they were flying their cars. That meant making the parapet as transparent and visually delicate as possible to allow your eye to take in the vast sweep of the valley. The bridge also curves gently on plan, on a 20km radius, so that when you are at the wheel of your car you are aware of the entire structure as you drive across it.

Building the bridge was an epic endeavour. Typically, before assembling a structure such as this all the sections would be laid out below their final resting place and then hoisted into position. However, the contractor Eiffage decided that it would be much more cost effective to construct the bridge up in the air, with a minimum of temporary support towers.

This meant launching the road deck in two long sections from the plateaux on either side of the gorge to meet at a point above the river. This was an audacious choice and was to involve the world's longest ever bridge-deck launch, requiring a specially designed set of computer-controlled, synchronised horizontal jacks on each of the permanent and temporary supports.

Today, we tend to take the technology involved in building the bridge for granted. But it was interesting to note that, as little as a decade earlier, a project of this scale and ambition would have been almost impossible within the budget and timescale we set ourselves. Without the special computer-guided machines, which were used to cut and assemble the various structural components, or the guidance of satellite fixes, the margin of error in laying out the carriageway – in the order of plus or minus 3mm – would simply not have been achievable.

When I went to see the bridge as it began to take its final shape I was naturally apprehensive. Had we made the right choices? However much one models or draws something, one can never be quite sure how it will look in reality. When I saw the bridge I had a huge sense of relief. It is very elegant, very taut, and creates a dramatic silhouette against the sky. I was also pleased to discover that local reaction to the bridge as it took shape was hugely positive.

On 16 December 2004 – one month ahead of schedule – the Millau Viaduct opened to traffic. A process that had begun in 1987, with preliminary feasibility studies, was complete. The A75 now provides the shortest, high-speed route from Paris to the Mediterranean coast and on to Barcelona. On 22 July 2011 another major landmark in the bridge's history was reached when the 30-millionth vehicle completed the crossing. That equates to an astonishing average of 12,500 vehicles per day.

Not only has the bridge relieved Millau of traffic, it has also provided the town with a new landmark – one hears stories of people going out of their way simply to drive across it. That degree of popularity has brought additional benefits. A town once in economic decline has found new life as a tourist destination. Freed from gridlock – and all the environmental problems associated with it – Millau has been rediscovered as a place of great natural charm.

When the viaduct was conceived, I do not believe that the social and economic benefits that can accrue from well-conceived infrastructure were at the front of people's minds. But the results speak for themselves. Considering how controversial this project once was, that conclusion is extremely gratifying for every one of us associated with the project.

The architecture of infrastructure Thomas Leslie

Millau Viaduct is the result of a creative collaboration between architects and engineers and the culmination of eleven years of design and construction work. More significantly, perhaps, it confirms Norman Foster's belief that an approach which emphasises the synthesis of structure and programme into a unified, coherent form has validity across a range of scales – even when applied to such a daunting undertaking as a 2.46km-long, sky-high motorway bridge.

Major bridges represent extreme territory for design and engineering in that the combination of exceptionally long spans, stringent demands for safety and performance, and the inevitable demands of public works budgets place unusual constraints on form and expression. As the Chicago architect-engineer Myron Goldsmith suggested in 'The Effects of Scale',[1] the limitations and economics of various systems determine structural forms and approaches at different spans and conditions. While an arch might be the best solution for a span of 100 metres, for example, the challenge of spanning 1,000 metres can today only be met by a suspension system of some kind.

Thus the bridge designer begins with a structure suggested to a large degree by the conditions of the site and the problem to be solved. At first sight, there would seem to be few opportunities for the sort of freewheeling exploration and lateral thinking for which Foster is so well known. Indeed, bridge design has become a specialty even within engineering, and architects have traditionally been excluded from such a focused discipline.

Worse, when able to participate in such projects, the architect is often limited to decorating an otherwise resolved scheme. Perhaps the most notorious example of this is Othmar Ammann's George Washington Bridge in New York, completed in 1931. Cass Gilbert – architect of the Woolworth Building in Manhattan – was commissioned to give the bridge a pair of granite-clad towers. The Depression forced the abandonment of his scheme, leaving the towers' raw steelwork exposed and forming an iconic engineering statement that impressed even Le Corbusier. In hindsight, Gilbert's massive pylons would have been almost tragic, given the visual richness of what lay beneath, and this near-miss has long stood as an example of the problems inherent in trying to integrate engineering principles with architectural style.

There is, nonetheless, a long history of bridges that have teased iconic form from the seemingly onerous requirements of span and structure, particularly in France. For example, Gustave Eiffel's Viaduc de Garabit, which opened in 1885, springs

Previous pages: A cutaway drawing of the bridge deck. The deck rests on the concrete piers via spherical bearings. The steel deck structure took the form of a 'kit', with units no more than 22 metres long, to allow them to be brought to site by road.

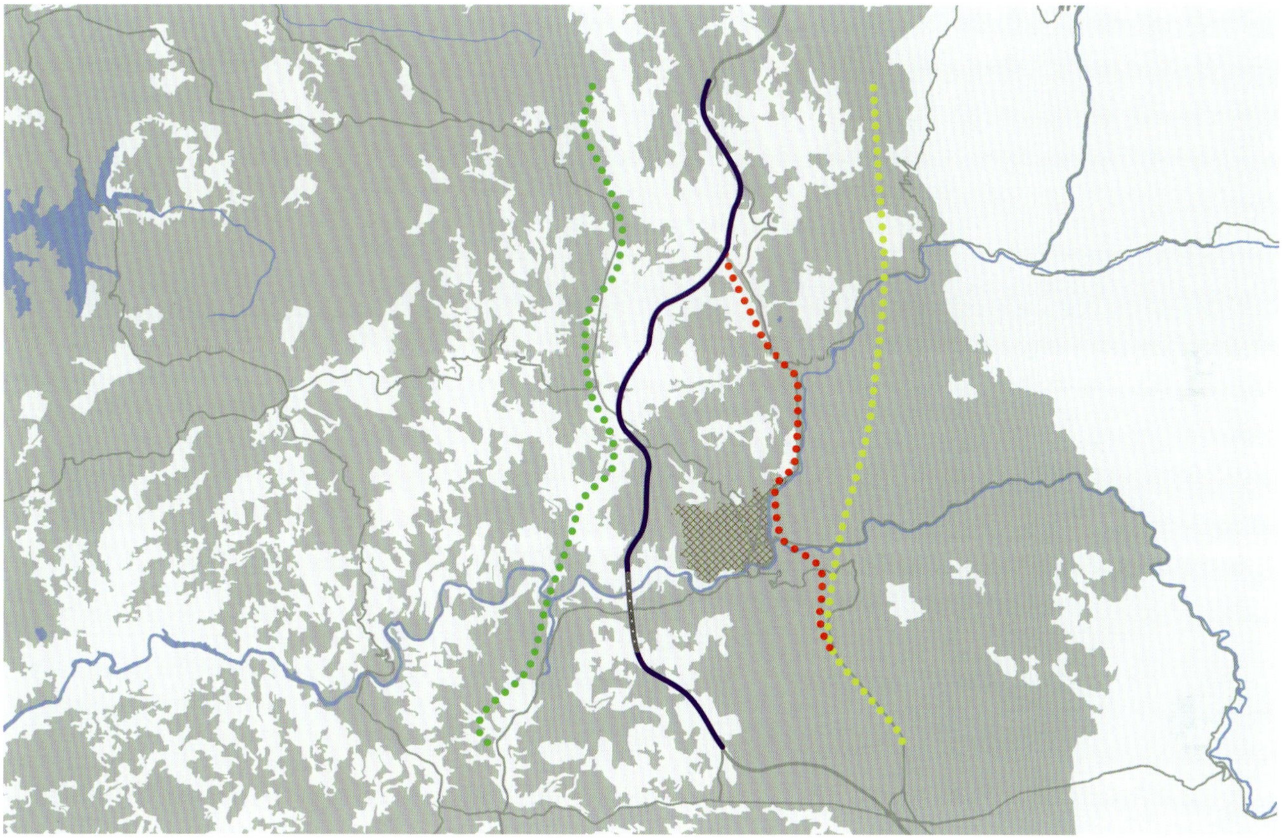

Far left, left: Before the bridge was built, motorists were forced to follow a narrow, meandering route across the valley floor; traffic congestion in Millau itself was a major problem in the summer months.

Above: A map showing the location of the bridge and the route of the A75 motorway. Also shown are the four routing options – respectively, red, yellow, green and blue – across the Tarn Gorge, which were examined in detail in 1987. The blue route was approved in 1989.

Right: A model of the initial competition scheme for the bridge, which proposed an under-slung catenary system, supporting 600-metre spans.

across its valley site with a tapering, parabolic steel arch that is both structurally logical and visually expressive. Twentieth-century designers, such as Robert Maillart, explored the static grammar of concrete spans to reinforce a similar narrative of performance and construction. Maillart's elegantly minimal Swiss Salginatobel and Schwandbach bridges, from the early 1930s, have had a profound influence on the aesthetics and engineering of contemporary bridge construction.

While steel inevitably reflects a story of components combined to create an efficient whole, monolithic concrete suggests designs that stress continuity and single statements. Architecturally successful bridges are generally those that tend toward refinement and purification, teasing out and celebrating the static and constructive principles of the engineering scheme.

At the time of the Millau Viaduct competition in 1993, Norman Foster was not known for bridge design, though the practice had participated successfully in a competition for a light-rail viaduct for Rennes, in Brittany, and its work on projects such as Chek Lap Kok Airport in Hong Kong (1992-1998) and the Bilbao Metro (1988-1995) had given the firm extensive experience in transport infrastructure and engineering. More immediately, Foster's Torre de Collserola in Barcelona (1988-1992) gained worldwide acclaim as a successful fusion of architectural engineering principles – exactly the challenge presented by the bridge.

Millau Viaduct was proposed as the final link in the A75 – the motorway that links Paris with the Mediterranean coast and then on to Barcelona. The motorway at the time discharged traffic on to a narrow route nationale that passed through the town of Millau, creating a bottleneck that in high season could result in delays of several hours. Bypassing Millau was a vital step in establishing better road links between France and Spain. However, the site in the Massif Central presented a classic French landscape and inevitably there were concerns about the impact of such a project on views from Millau along the Tarn Gorge. The unloved Pont de Saint-Nazaire (1975), across the Loire, served as a warning of what might result in Millau if insufficient attention was paid to the viaduct's design. The French authorities shrewdly decided to allay such fears by inviting architects and engineers to enter an open competition.

Foster entered the competition teamed with Arup. At that time, the practice maintained an office for the Carré d'Art, in Nîmes, just two hours' drive from Millau,

Left: Alistair Lenczner at one of the regular meetings held on site with the team from Compagnie Eiffage du Viaduc de Millau, discussing the detail of the top of the mast, which includes aircraft warning lights.

Left: The Pont de Normandie, over the Seine estuary in northern France, designed by Michel Virlogeux and constructed between 1988 and 1995. When completed, it was the longest cable-stayed bridge in the world.

and so Foster and his colleagues knew the area well. The competition organiser, Service d'Etudes Techniques des Routes et Autoroutes (Sétra) – the French department of transport and public works – had presented entrants with their own initial structural concepts, and had concluded that a multiple cable-stayed system, with masts and cable fans, would provide the most economical solution. Sétra's basic layout – devised by the French engineer, Michel Virlogeux – consisted of a phalanx of seven pylons across the gorge, each pylon carrying a fan of cables to support the bridge deck.

Foster's initial competition entry included a fresh study of how the gorge might be spanned. Working with the late Tony Fitzpatrick of Arup, a team in the London office devised a scheme that relied on under-slung catenary cables to support 600-metre spans, thus reducing the number of supporting towers while leaving views from the roadway unencumbered by cables or towers. The result was so lightweight that uplift from wind presented as serious a structural issue as the weight of the deck itself, and the resulting webs of supporting and tie-down cables provided what would have been a unique construction across the valley.

In addition to the under-slung cable scheme, the Foster team provided a detailed analysis of the Sétra

Above: Michel Virlogeux and Norman Foster standing on the Causse Rouge plateau at Millau on 16 October 2003. At that stage, the two road decks were in the process of being launched from either side of the valley.

Right: The Millennium Bridge, London (1996-2000). The result of a unique collaboration between an artist, an architect and an engineer, it was the first of Foster + Partners' bridge projects to be completed and the first new Thames crossing in more than a century.

layout, noting that the proposed towers seemed ill-proportioned and that there were alternatives to their spacing that would harmonise better with the overall appearance of the bridge from below.

Ultimately, the under-slung scheme did not prove to be as economical as the cable-stayed solution, but the competition jury was impressed by the team's design agility, and Foster was asked to submit to the second stage of the competition, developing the cable-stayed scheme in conjunction with French engineers. In retrospect it is clear that, in part, Foster was awarded the commission on the basis of his willingness to question Sétra's assumptions and to offer alternatives.

Interestingly, the viaduct proved to be the precursor of a number of high-profile bridge projects on which Foster would work in the 1990s, most notably the Årsta Bridge in Stockholm (1994-2005) – a supremely elegant and understated railway viaduct – and the Millennium Bridge in London (1996-2000). The Millennium Bridge – what Foster has called the 'Blade of Light' – is particularly significant in that it sprang from a collaboration between architecture, art and engineering, with sculptor Anthony Caro and engineers Arup. The only dedicated pedestrian bridge in central London, and the first new Thames crossing in more than a century, it represents a powerfully forged architectural and urban statement.

From the outset of the Millau competition, the Foster team found themselves immersed in the extreme pressures of bridge design. While the studio is more used to examining solutions from first principles, here the challenge was to work within a framework that was in some ways predetermined: the brief was to develop and refine. Alistair Lenczner, an engineer by training who led the Foster team from 2001, observes that there is always more than one way to explore an engineering solution: 'Sometimes you have to stop and question whether something cannot be done better; if something looks and *feels* better it usually works better as well.' Ultimately, this was to become Foster's most important contribution: honing and articulating the formal logic of the engineering solution and distilling it into a clear aesthetic statement.

Millau Viaduct is among the largest projects ever addressed by Foster, yet its design methodology bears similarities not with that of other large buildings, but with some of the smallest industrial-design projects. In product design, compelling form emerges from a functional and sculptural sensibility and the pure logic of fabrication and use. Tim Quick, who led the Foster team until 2001, recalls that while he was working on

Millau he sat across from an industrial designer who was working on a range of bathroom fittings: 'Curiously, we found we had more in common with each other than with the rest of the office, because we were dealing not with occupied space but with an object that's seen in the round.'

Like a basin or a door handle, the programme for a bridge is in most respects very simple. Millau states that programme – 'from here to there' – eloquently and economically. There is no redundancy, no grand gestures, only the refinement of a logical solution coupled with the search for an elegant, communicative form.

The project was fortunate to have had, in Michel Virlogeux, a brilliant engineer who understood the importance of the architect's contribution to bridge design. Virlogeux had not only been instrumental in devising Sétra's preferred structural concept, he also welcomed the refinement he knew Foster would bring to the scheme.

Having left Sétra, Virlogeux 'decided to join the team who I believed were developing the project in the best way. And so I just approached Foster and said "If you don't mind, can I join your team?" "Sure", he said.' That relationship was to prove highly productive. Virlogeux and Foster had much in common, not least a belief that 'the best things often result from the simplest ideas.'

Foster observes that 'during the design process, there were never any conflicts between satisfying the structural demands and our aesthetic ideas: they evolved together.' However, his instinct as a designer is constantly to question and explore alternatives and so naturally he wanted to review all the engineering decisions from a conceptual and experiential point of view. Much of this had to do with Foster's concern that such a dramatic application of static principles be made legible.

Foster and his team focused on keeping the scheme as 'clean' as possible – eliminating or downplaying elements that would distract from the 'pure' experience of the structure. His first move was to suggest a change to the spacing of the masts in order to spread the cable fans out to the very ends of the deck, thus eliminating what would have been a weak-looking span at either end of the bridge.

As the project developed, it was Michel Virlogeux who prompted a solution to a fundamental structural issue. Because concrete and steel expand and contract as the temperature changes, the road deck had to be designed to move seasonally by up to a metre at each end. Conventional practice dictated that movement be

Left: The Årsta Bridge, Stockholm, completed within a year of the Millau Viaduct, in 2005. A railway bridge, which incorporates pedestrian and cycle paths, it is designed to complement the historic Årstabrun Bridge dating from 1929.

Right: One of the many study models made to explore the junction between the road deck and the supporting piers.

allowed for at each pier to allow the expansion of each successive span. However, this would have entailed seven sets of maintenance-intensive expansion joints. Instead, Virlogeux proposed making the deck continuous, taking up the whole expansion in two massive joints at either end of the bridge.

To achieve this, a device was introduced to provide the concrete piers with the required bending strength while giving them more flexibility to accommodate the imposed temperature movements of the deck. This device involved splitting the piers into two separate elements at deck level, each of which can flex significantly more than a single, monolithic pier. The steel masts above the deck are also bifurcated, but connected at the top, thus maintaining their full carrying performance, but allowing the bridge at road level to be more 'supple'.

The massive forces that result from the bridge's dead load and the impact of wind funnelled along the length of the gorge means that the piers are in themselves gigantic structures – Pier 2 at 244.96 metres, is the tallest ever built.

The original Sétra scheme had proposed piers with a rectangular cross-section. Again this was challenged by the Foster team. They worked with the engineers to devise an elegant, faceted form, which tends to catch the light and reduce the visual weight of the piers against the landscape. Many models were made to test and 'distil' this complex form until it was judged just right.

Each pier forms a hollow tube, composed of a concrete skin that tapers and folds as it rises to provide additional stiffness. Any extra costs deriving from the complexity of the concrete formwork were more than compensated by efficiencies in the structural performance. The folds also provide visual continuity, allowing the 'eye' of the mast to be integrated with the tapered form of the pier to create a distinctive 'needle'.

The seven 'needles' instantly give the bridge iconic form. Foster has also likened the piers to blades of grass or butterflies' legs for their slender proportions. Further design development resulted from a concern for the piers' connection to the land. Bridges of this scale are necessarily invasive, as the towers require massive foundations. Foster was adamant that the piers should give the impression of rising from the earth naturally and reaching skyward – like giant redwoods. He therefore insisted that the pier foundations be buried deeply and that the surrounding topography and flora be restored after completion. Specifications were drawn up that guaranteed the restoration of the site to its original state following

Left: Short and tall versions of the mast were tested, as a single design solution was sought that would work across all seven mast conditions.

During the design process, there were never any conflicts between satisfying the structural demands and our aesthetic ideas: they evolved together.
Norman Foster, 2011

Above: A sketch model of the final scheme. To accommodate the expansion and contraction of the steel deck, each concrete pier splits into two thinner sections at a point 90 metres below the roadway. The steel masts form A-frames which carry eleven pairs of cables to support the deck.

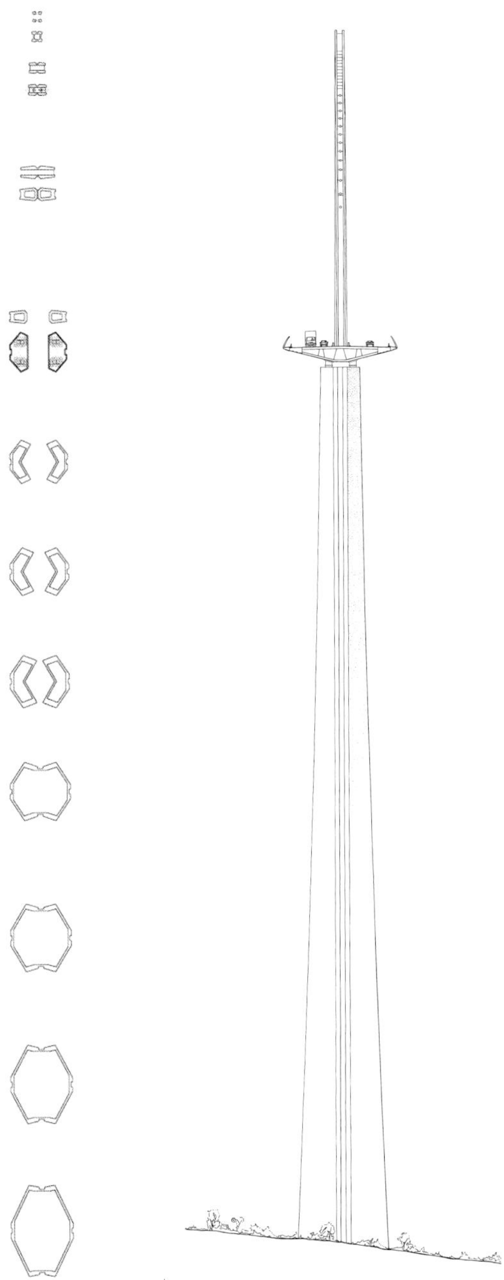

construction, highlighting the dialogue between the rough landscape and the finely honed object traversing it. Strict provisions were also made for water treatment, dust control and soil remediation, ensuring that the massive construction operation made as small an impact on its surroundings as possible.

Furthermore, all subsidiary elements – electricity transformers and so forth – are accommodated within the footprint of the piers, and maintenance access lifts and stairs are concealed, thus avoiding the forest of ladders and catwalks that typically clutter large bridge projects. As a result, the viaduct sits with almost impossible visual lightness upon the slopes of the gorge, to the point where it is virtually a work of land art in its own right.

The original arrangement of the cables was refined to allow a single row of cable fans centred on the deck. This solution contrasts with more typical cable-stayed designs, which commonly provide a pair of fans for each mast, connected to the outer edges of the deck. The rationale here was that from a distance the cables would have created distracting interference patterns between the angled fans, and from the driver's perspective would have obscured the view of the gorge. To maintain structural stability, the deck was redesigned as a closed box-girder section, which

Left: A cross-section through the road deck, looking towards Pier 2, and the corresponding sections through the concrete pier and steel mast; there are seven piers, spaced at 342-metre centres.

Right: Pier 2 and the road deck seen in elevation. The roadway has a constant slope of 3.025 per cent from north to south.

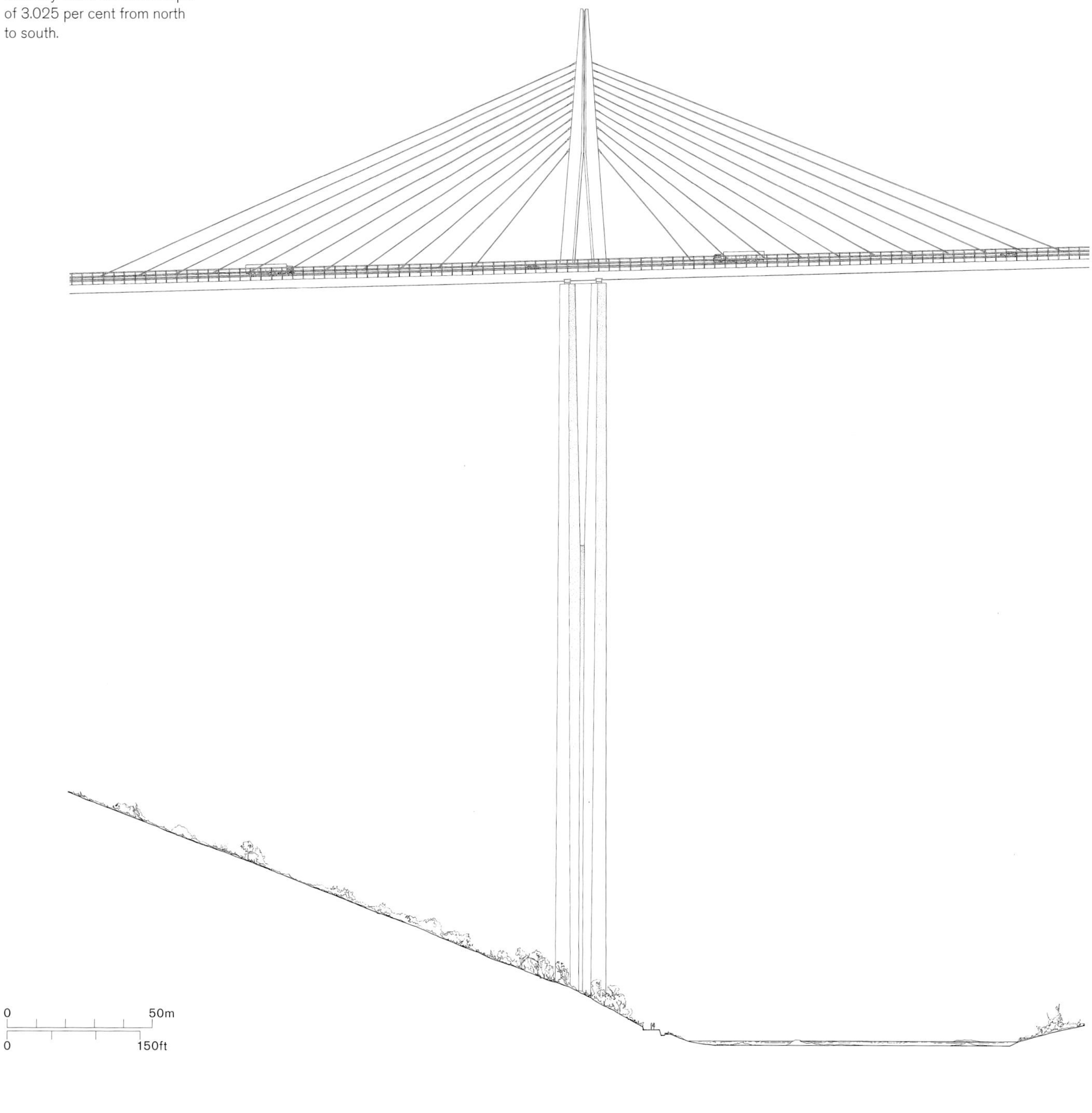

We wanted the piers to look as if they had barely alighted on the landscape, light and delicate – like butterflies' legs. Norman Foster, 2011

allows it to transfer torsional forces into the masts efficiently; the slender profile of the leading edge – akin to that of an aircraft wing – mediates its central depth. In the process, the number of cables and connections was reduced by half, thus improving the image and experience of the bridge. The cable connections were designed with vandal-resistant shrouds, which conceal the cable-end fittings and provide a smooth detail appropriate to the speed of passing traffic.

Foster also suggested a gentle curve to the deck on plan, giving motorists a slightly oblique view of the seven cable fans as they arrive at the springing point of the deck. This geometry avoids the end-on approach – common in bridge design – which prevents drivers from appreciating the linear progression of structure along the span. It has the added benefit of alleviating the gradient between the two sides of the valley, of 3.025 per cent from south to north.

The way the bridge's superstructure is seen against the sky led Foster to decisions about colour. Generally, to make an object recede in the landscape, you make it dark. But here the bridge is seen mostly against the sky and so Foster decided that the superstructure should be painted in light shades – the masts and deck are pale grey and the cable shrouds white.

Left: The design was developed in a way that allowed the bridge deck to be built either in concrete or steel. The concrete deck was heavier and required more cable stays, as shown in this model.

Left: Norman Foster, Mouzhan Majidi and Alistair Lenczner seen in the Foster studio at Riverside, in London, with a model of the final scheme.

This means that what the engineers call the 'skyworks' offer a delicate counterpoint to the blue of fair weather skies or disappear ethereally into the clouds on overcast days.

Perhaps the most critical design element from the driver's point of view was the form of the aerodynamic side screens, which protect vehicles from strong cross winds. The Foster team developed a screen composed of horizontal acrylic vanes, which are attached to the deck by means of curved steel stanchions, which spring from the edge of the deck. The screen is detailed in such a way that it is high enough to cocoon the roadway and yet sufficiently transparent to maintain views out across the valley. The strong horizontal emphasis also tends to emphasise the sense of speed that comes naturally on a highway at such an extreme elevation. By making the side screens partly open to air movement, the strength of turbulent eddies over the top of the screen was reduced.

In August 2001, Compagnie Eiffage du Viaduc de Millau (CEVM), a subsidiary of the French construction company Eiffage, was awarded the contract to build and run the bridge. In exchange for a seventy-five-year concession, Eiffage agreed to meet the entire cost of construction, which would be recovered through toll charges. Management of the detailed construction design process was handled by Setec TPI, an independent engineering contractor, and subcontracts were let to a wholly French cast of builders and suppliers, including Eiffel for the steelwork and Freyssinet for the cabling – two distinguished names in bridge-building history.

The tender process highlighted decisions about the choice of materials, thus prompting a further refinement of the scheme. The design team had allowed for the bridge to be built either in steel or concrete, as the client wanted to allow for both options to ensure maximum competitiveness. French steel construction in the late 1990s had generally been prohibitively expensive, but changing construction economics meant that steel was suddenly affordable.

In preparing their tender, Eiffage determined that both the road deck and the masts above deck level should be built from steel. This decision allowed the number of cables to be reduced still further, as a steel deck is significantly lighter than its concrete equivalent. It also meant that the vocabulary of the structure changed subtly. The continuity of the concrete would have suggested much more strongly a bridge of two components: pylons and roadway. The introduction of steel for the masts now had the potential to introduce a third set of elements to the visual hierarchy.

Right: The model was taken up to Dunstable Downs in the Chiltern Hills and photographed against the backdrop of a real sky.

Above: On 9 December 2003 more than 400 people attended a spectacular firework display, to celebrate the completion of the concrete piers.

Left: Norman Foster on a visit to the Millau site in July 2003; the southern deck, which was launched in March, awaits the next stage in the procedure.

From an intellectual standpoint, one might expect the legibility of the structure to be compromised. However, in reality the bridge is experienced in one of two quite different ways: from a distance in the landscape, or close-to from the road – there is no middle ground. From a distance, in silhouette, the change in materials cannot be discerned – the masts and piers read naturally as one – while to the driver's eye, the bridge is entirely of steel, so there is no inconsistency. Occasionally, when fog fills the valley, the deck appears to rest on nothing but vapour – a surreal if unintentional effect that further emphasises the lightness of the superstructure.

Steel ultimately won over concrete for the deck and masts because it allowed for a rapid construction process once the piers were completed. There was no practicable way to pour a continuous concrete ribbon at the height of the deck without extensive scaffolding and multiple cranes, while the lighter steel deck could be prefabricated and 'launched' progressively from the two abutments with the minimum temporary support.

Eiffage devised a technically advanced system that allowed a continuous steel box-girder deck to be launched across the tops of the seven concrete piers and seven intermediate temporary steel towers, the tallest of which was 173 metres high. The intermediate towers were intended to limit the unsupported spans and thus the temporary bending forces in the deck as it was launched without the support of the cable stays. Like the permanent piers, each intermediate tower was equipped with computer-controlled hydraulic jacks, which moved the deck incrementally. Because it was impractical to place a 300-metre-tall temporary support where the deck crossed the Tarn itself, the leading sections of deck launched from the north and south abutments had to include a mast and cable stays to provide support for the 'flying' cantilever ends.

As the two leading sections of deck advanced, the structure behind was welded together – using the road surface on either side of the gorge as an assembly area – then pushed out as a continuous box girder, moving a maximum distance of 600mm every four minutes. In total, 1,540 push cycles from the north and 3,280 from the south were required to complete the entire span. When the final welds between Piers 2 and 3 were completed, the result was probably the longest box girder of its kind in the world.

This construction methodology was highly efficient in that it made maximum use of prefabrication and rapid in-situ assembly, and limited the amount of material stored on site. Individual components were

Right: On 7 April 2004, while the Queen and Prince Philip were on a state visit to France to mark the centennial of the Entente Cordiale between Britain and France, the Duke visited Millau and met members of the design and construction teams.

Right: In 2005 to celebrate the opening of the Millau Viaduct, the French postal service, La Poste, issued a special 0.5 Euro stamp, designed by Sarah Lazarevic.

sized to be deliverable by trucks small enough to negotiate the region's narrow roads.

With the deck in place, the next work phase was to install the steel masts, each of which weighs 700 tonnes. The masts were brought to site in sections and assembled on land before being moved into position and erected, using a system inspired by the hoist developed in ancient Egypt to raise obelisks. Each mast carries two sets of eleven stays, comprising 55 to 91 high-tensile-steel strands, depending on their length. The final operation was to clean the road deck and lay a 70mm layer of asphalt, an operation undertaken by Appia, Eiffage's road-building subsidiary.

Discussions about signage and roadway lighting revealed that neither of these elements was necessary on the bridge – the toll booths are located away to the north, and night-time traffic arrives at the viaduct after several kilometres of dark, rural motorway. Both were thus eliminated, and with them the potential distraction and clutter typical of most bridges. However, to give the bridge a night-time presence from Millau itself, a subtle lighting scheme for the masts was devised in collaboration with Claude Engle, a regular collaborator with the Foster studio. Other elements required on the bridge – emergency telephones, video surveillance cameras and weather stations – were integrated into the masts or rails, subsuming them into the overall scheme.

Millau Viaduct opened on 16 December 2004 – a month earlier than planned – to local and global acclaim. Speaking at the opening ceremony, President Jacques Chirac declared the viaduct 'a magnificent example, in the long and great French tradition, of audacious works of art – a tradition begun at the turn of the nineteenth and twentieth centuries by the great Gustave Eiffel.'

Press reports revelled in the astonishing statistics that are requisite for any major bridge. Millau Viaduct has the highest road bridge deck in Europe; its construction involved the world's longest high-precision bridge-deck launch; it has the highest piers in the world, breaking the record held by the Kochertal Viaduct in Germany; it has the highest tower in the world – the mast atop Pier 2 reaching 343 metres (beating the Akashi Kaikyo bridge in Japan – the world's longest span suspension bridge); and perhaps most significantly, it displaced the Eiffel Tower as France's tallest structure. However, as has been said of the Empire State Building, Millau Viaduct may be famous for being big, but it should rightly be famous for being good.

Left: Another spectacular firework display marked the opening of the viaduct on 16 December 2004. It was a celebration but also a farewell from the 537 people that had worked to complete the structure.

The one-minute drive across the bridge is a sublime automotive moment, matched only perhaps by the journey over San Francisco's Golden Gate Bridge. On the left you experience the staccato rhythm of the seven towers and the fine grain of the cable fans, while to your right is the shifting panorama of the Tarn valley. The entire sequence is visually composed – there are no fussy details or signs to compromise the ordered experience of the crossing – and the clarity and formal power that one perceives from a distance remain consistent close to.

Millau Viaduct has proven unique in its drawing power, particularly considering its remote location. No tourist amenities were planned for the base, but the sight of towers the size of skyscrapers marching across the landscape has become an irresistible lure for visitors to the town – and to those curious to see the scale of the bridge from a different perspective. There is surely an irony in drivers making a lengthy detour to see a viaduct expressly designed to quicken their journey.

'If you care about the place to which you bring the bridge', says Foster, 'then it follows that it should be elegantly conceived – that it should have the ability to inspire.' These are the kinds of benefit that cannot be accounted for conventionally in spreadsheets. The elevation of a heroic work of engineering to a singular, memorable automotive experience has brought fame and even tourist revenue to Millau even as it has set a new standard for highway design in France and elsewhere. As with the Millennium Bridge, the Millau Viaduct has generated economic benefits in its locale and beyond, which over time will far exceed the investment in the project itself.

It may be that the overwhelming success of the Millau Viaduct – aesthetically, functionally and economically – will serve as a model for future large-scale projects. It has demonstrated the value of the design vision that can result when architects and engineers work together in an integrated way, even in projects of such formidable size. The viaduct's legacy therefore extends far beyond its impact on the transport infrastructure of the region, and even the model it offers for bridge design. It represents a bold departure from traditional disciplinary limits, showing how architecture has a fundamental role to play in developing basic engineering principles into legible, meaningful experiences.

1. *Myron Goldsmith: Buildings and Concepts*, edited by Werner Blaser, 1987

Above left: Norman Foster with the President of the French Republic, Monsieur Jacques Chirac, shortly before the opening ceremony on 14 December 2004.

Above right: At 11.15am, President Chirac unveiled a plaque commemorating the completion of the Millau Viaduct. The bridge was opened to traffic a month earlier than originally planned to allow people to drive across it on their way home for Christmas.

Left: A billboard announcing the opening of the bridge.

Above: An aerial view of the bridge, seen from the west. The incredible scale of the endeavour and the delicacy of the structure are both apparent when viewed from the air.

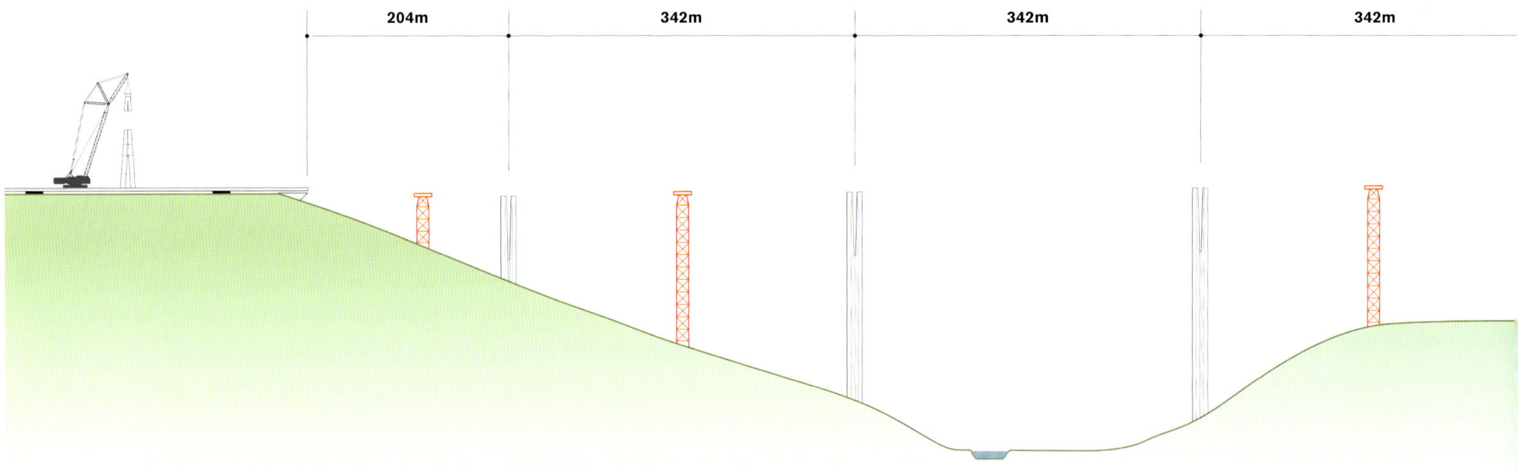

Above: The bridge abutments and piers were the first elements to be built. There are seven concrete piers, spaced at 342 metre centres. Intermediate temporary steel towers were designed to provide support for the deck during the launch process – the next stage in the erection sequence.

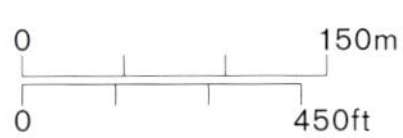

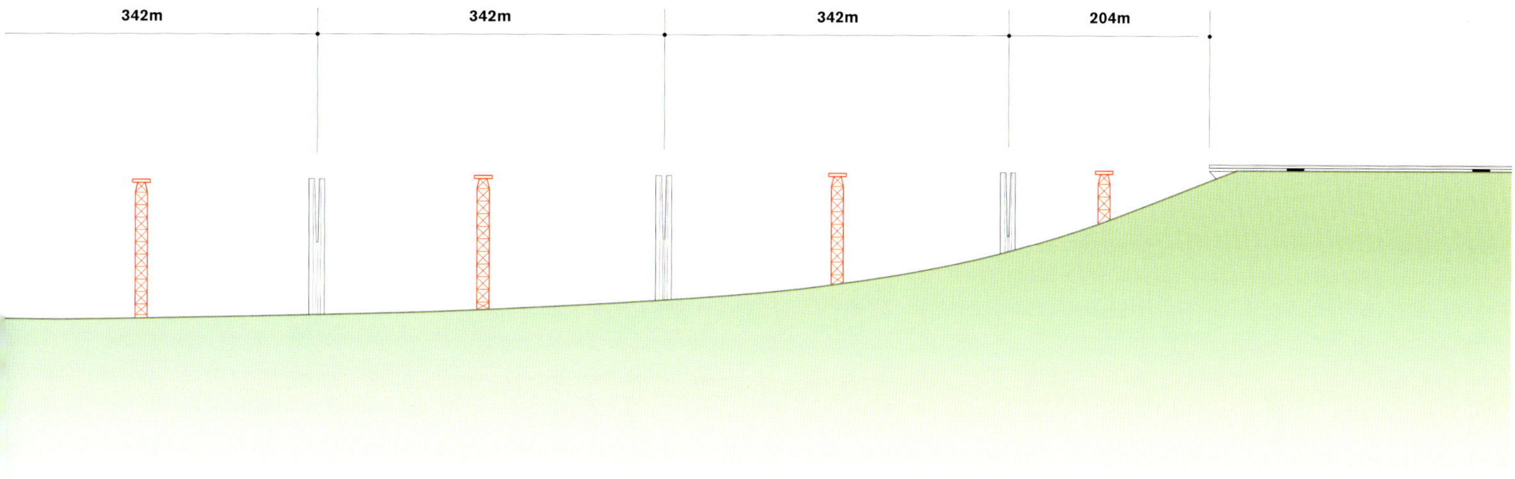

Overleaf: The viaduct was constructed by Eiffage – a descendant of the company that built the Eiffel Tower. The contract was awarded in October 2001 and construction took 1,096 days to complete. Here the seven work sites for the piers indicate the future alignment of the structure.

Above: Work on the pier foundations began on 4 February 2002; by 24 November 2003 the last of the concrete piers (P3) was complete.

Right: Each pier was treated as an independent site, with its own construction crew. The piers rose in height at the rate of one 4-metre pouring every three days. The geometry of the piers varied from one pouring step to the next. This required constant adaptation of the formwork, which was specially developed for the project.

We were driven by the scale of the idea and a shared passion for the poetic dimension of engineering and its sculptural potential. Norman Foster, 2011

Above: The first section of deck was launched from the south abutment on 25 March 2003. When completed, it was the world's longest ever bridge deck launch and involved the use of the first synchronised set of computer-controlled horizontal jacks.

Right: A 50-metre long steel 'nose' structure at the leading edge of the deck guided progress and facilitated docking on to the successive supports.

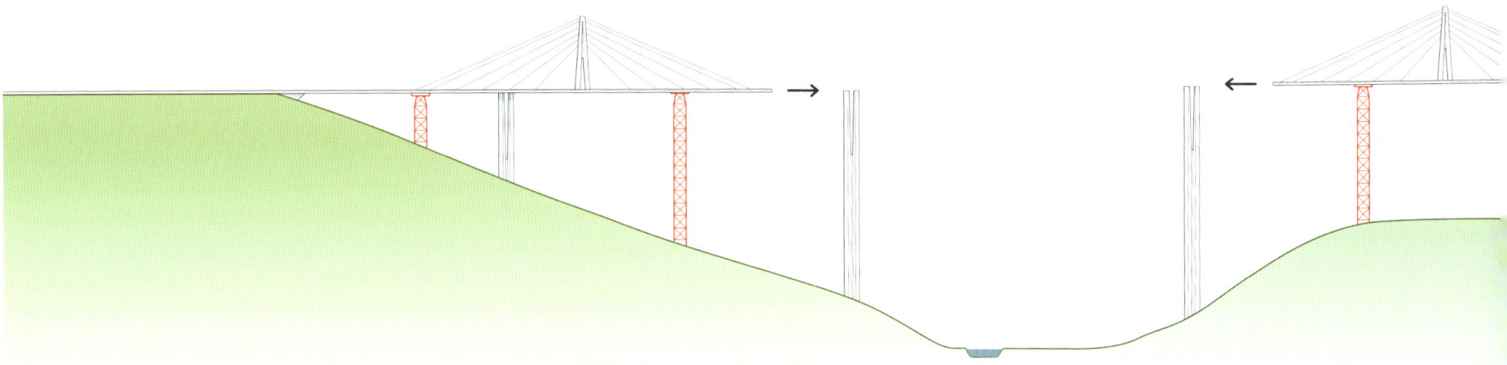

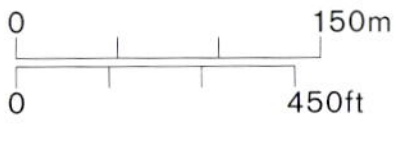

Above: A technically advanced system was designed to move the steel deck into position. The first stage involved moving the two central deck sections, which were launched from opposite sides of the gorge. Each of the seven temporary piers was equipped with a hydraulic system to aid this process.

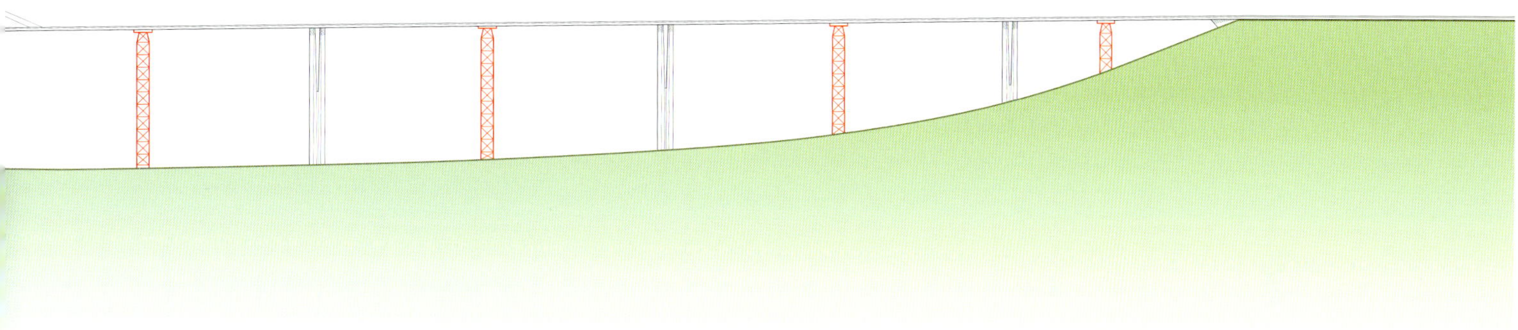

Above: The first sections of deck were launched with their 87-metre steel masts in place to support the cantilevered leading edge.

Right: At the launch stage, only twelve of the final twenty-two cables that support the deck were attached to the masts.

The greatest art comes from making things very simple, but very elegant and perfectly adapted. Michel Virlogeux, *New Civil Engineer*, 10 February 2005

Above: The leading edge of the southern deck approaches the first of the temporary intermediate piers; March 2003. The tallest of the seven temporary piers was 173 metres high.

Right: The southern deck moves towards the first pier; during this procedure the deck advanced at an average speed of 7 metres per hour.

Above and right: The northern and southern sections of deck inch closer together. The deck at this stage is strong enough to span between supports, but not rigid enough to span without sagging.

Above and right: The launching system was attached to a temporary framework at the top of each pier; a metal arm extended from one side to guide the deck as it advanced, guided with great precision by satellite.

Above: On 24 April 2004 the northern deck was finally in position at the junction point, waiting for the southern section to arrive.

Right: The two sections of deck edge closer together. At this point the road deck is now 270 metres above the River Tarn, making it the highest elevated roadway in Europe.

Millau is in the tradition of the great European viaducts by Robert Maillart and Christian Menn, but on an even larger scale. It is a triumph of modern technology and construction and an example of the best collaboration between engineering and architecture.
Tony Hunt, *Architectural Review*, June 2005

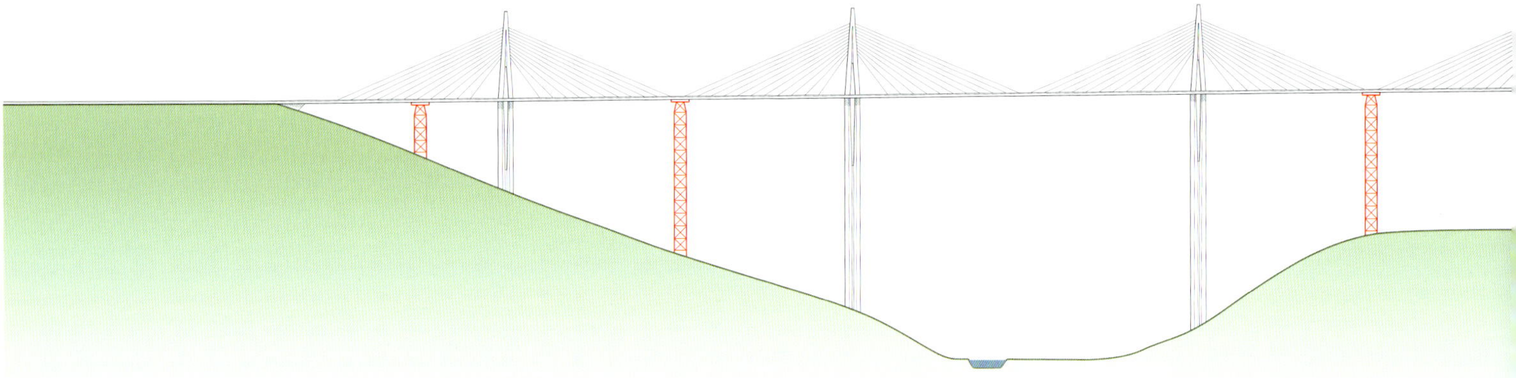

Above: With the deck in place, the remaining five masts and stay cables were erected. Once this process was completed, the temporary piers were removed, leaving seven needle-like pylons to support the weight of the deck. When the roadway had been surfaced, load trials and safety tests were carried out.

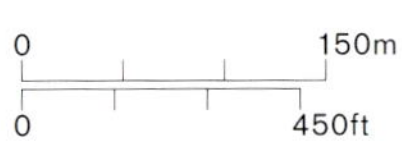

Without the satellite the viaduct would never have been built. The margin of error in laying out the carriageway was of the order of 3mm and its slightly undulating shape meant that the usual methods of calculation were useless. Marc Buonomo, Eiffel project director, quoted in *Abitare*, February 2005

Above: The masts were assembled on the ground then manoeuvred into position on the bridge deck.

Right: The mast above Pier 1 in the process of being righted. Using a system of hydraulic jacks, the erection process for each mast took eight hours. As the cables attached to each mast were tensioned, the sagging of the deck was gradually reduced until the roadway was level.

Overleaf: Three masts are in position; the mast above Pier 4 is in position on the deck waiting to be erected.

Left and above: The pylons were fabricated and delivered to site in sections. Assembled on land, they were moved into position then lifted using a system inspired by the hoist used to raise ancient Egyptian obelisks.

Overleaf: The seventh and final pylon is raised into position.

A project like this would have been impossible ten years ago. The bridge needed special machines to assemble its various structural components, as well as new materials, some of which have been used here for the first time.
Marc Buonomo, Eiffel project director, quoted in *Abitare*, February 2005

Left: Once the road deck was complete, the next step was to clean and prepare the surface ready to receive a 70mm layer of asphalt.

Above: The bridge structure is complete and the temporary steel towers are in the process of being dismantled.

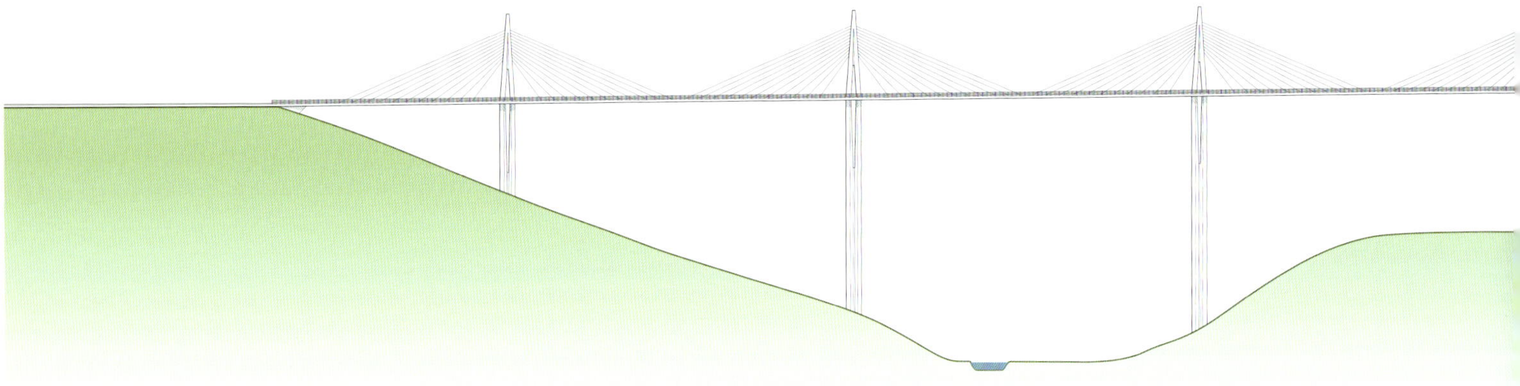

Above: With the deck in place, the remaining five masts and stay cables were erected. Once this process was completed, the temporary piers were removed, leaving seven needle-like pylons to support the weight of the deck. When the roadway had been surfaced, load trials and safety tests were carried out.

Right: Millau Viaduct compared with Gustave Eiffel's Pont du Garabit (1884-1888) and Eiffel Tower (1889). The mast atop Pier 2 peaks at 343 metres, 19 metres higher than the Eiffel Tower at 324 metres.

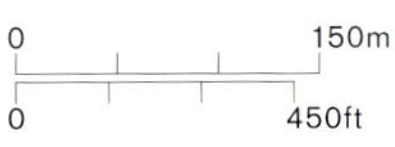

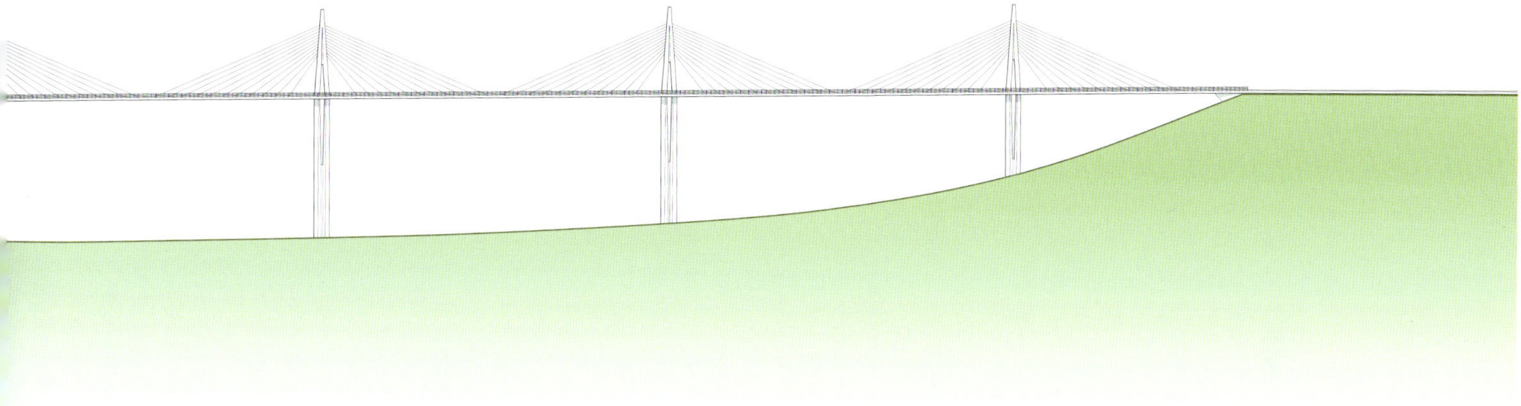

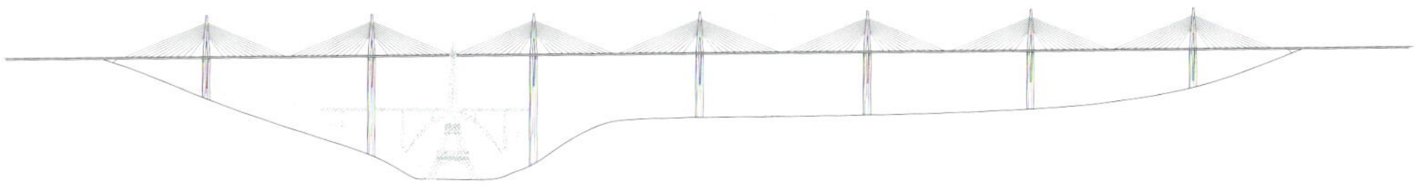

Overleaf: Seen from above, the 2.5km-long viaduct has a striking sculptural presence in the landscape. There is an interesting scale comparison with the historic bridge over the Tarn, seen lower left.

People go to Pisa to see the leaning tower and Paris to see the Eiffel Tower. We hope they will now come to Millau to see the viaduct. Jacques Godfrain, Mayor of Millau, quoted in *The Independent*, 25 October 2004

Left and above: The bridge seen from a vantage point high above the historic town of Millau.

The 2.46km-long viaduct spans the Tarn gorge to complete the A75 motorway from Clermont-Ferrand to Béziers across the Massif Central. The road deck curves gently in plan, on a 20km radius.

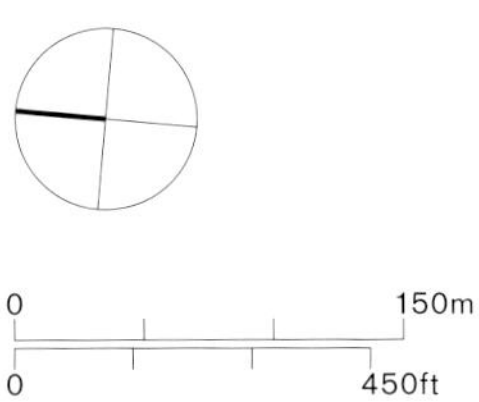

The curve in the bridge allows your eye to take in the vast sweep of the valley and you see the structure in its entirety ahead of you as you drive.
Norman Foster, 2011

The bridge seen from the west. It spans from the Causse Rouge plateau in the north – to the left in this picture – to the Causse du Larzac plateau in the south.

It took a feat of engineering and a leap of the imagination to span the rough, rugged Tarn Valley in the Midi-Pyrénées region of southern France. The result is breathtaking.
Elaine Sciolino, *International Herald Tribune*, 15 July 2005

With its trim cables carrying a sliver of a road 900 feet up in the sky, the bridge manages to be mighty and lissom at the same time. Editorial, *The Daily Telegraph*, 15 December 2004

Left and above: Contrasting perspectives: the suspension cables are painted white so that they read sculpturally against the landscape but recede when seen against the sky.

Above: Pier 7 is the southernmost and shortest pier, measuring 77.56 metres to the road deck, but even here the proportions are harmonious.

Right: Pier 2 – to the right in this view – is the highest pier and the tallest in the world, at 244.96 metres.

On a misty morning, from the valley's edge, the viaduct appears to glide through the clouds. But if the view from the ground is spectacular, drivers get nothing short of a widescreen masterpiece. Jonathan Bell, *Wallpaper*, December 2004

Above and right: Motorists often drive above the clouds, which form in the valley.

This may be the lightest, sleekest bridge design anywhere. Framed by the surroundings, it is pure environmental sculpture. Robert V Camuto, *The Washington Post*, 2 September 2007

Left: In this view from the north, the gentle curve of the bridge can readily be appreciated.

Above: Perceptions of the bridge change with the shifting light.

While the viaduct is undeniably dramatic, it is surprisingly graceful too. It doesn't so much dominate the gorge and plateau landscape as integrate with it. Andrew Mead, *The Architects' Journal*, 20 December 2004

Left: The 3-metre-high aerodynamic side screens are designed to shelter vehicles from very high winds while allowing views out to either side.

Above: Looking back along the south-bound carriageway.

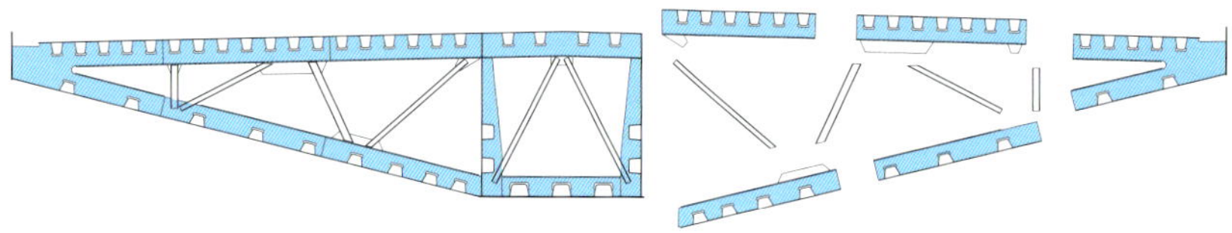

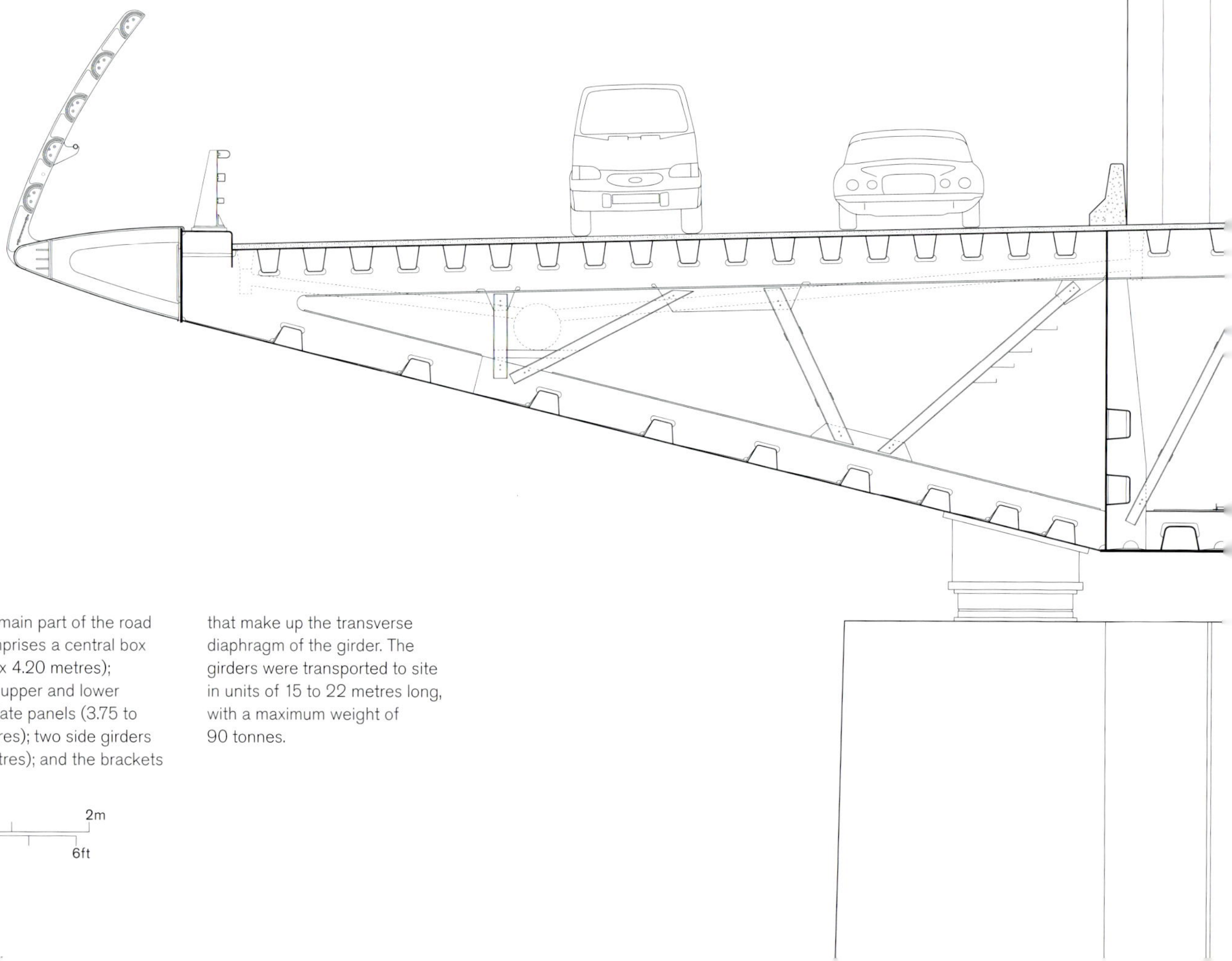

Top: The main part of the road deck comprises a central box girder (4 x 4.20 metres); stiffened upper and lower intermediate panels (3.75 to 4.20 metres); two side girders (3.84 metres); and the brackets that make up the transverse diaphragm of the girder. The girders were transported to site in units of 15 to 22 metres long, with a maximum weight of 90 tonnes.

0 2m
0 6ft

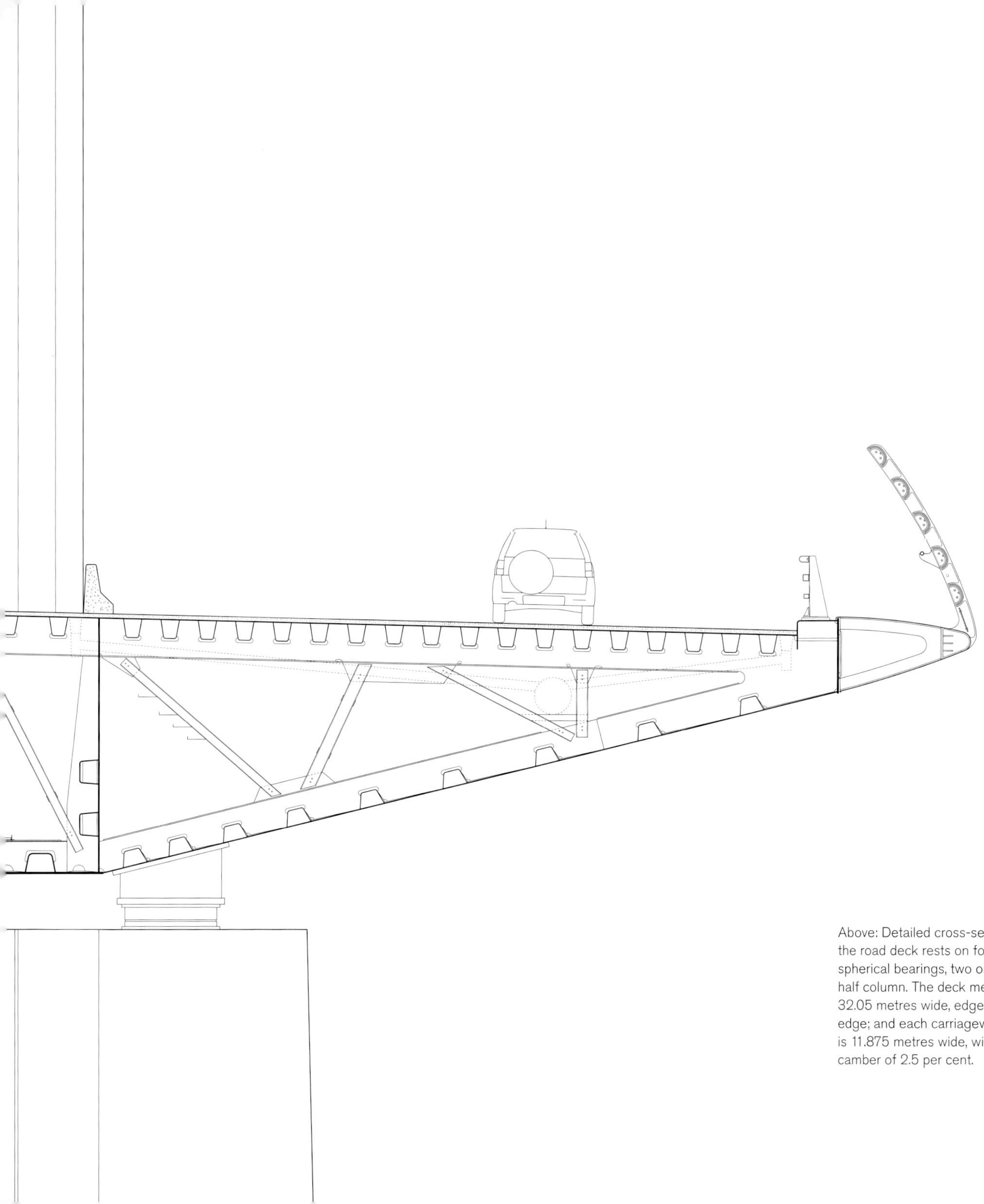

Above: Detailed cross-section; the road deck rests on four spherical bearings, two on each half column. The deck measures 32.05 metres wide, edge to edge; and each carriageway is 11.875 metres wide, with a camber of 2.5 per cent.

Left: Each column and mast is bifurcated to allow for the expansion of the road deck; this move also has the effect of lightening the appearance of the structure dramatically.

Above: Looking up at the tension cables that fan out on either side of each mast.

The bridge at night; the lighting scheme was designed to emphasise the upper part of the needles, using floodlights with very narrow beams.

None of us believed that the bridge would be finished so quickly, and none of us anticipated the response.
Guy Combret, director of Millau's Chamber of Commerce, quoted in *International Herald Tribune*, 15 July 2005

Facts and figures

Millau Viaduct

Millau, Aveyron, France
1996–2004

Client

Ministère de l'Equipement, des Transports, du Logement, du Tourisme et de la Mer

Project Team

Norman Foster

(Final scheme)
Tim Quick
Alistair Lenczner
Anne Fehrenbach

(Initial scheme)
Kevin Carrucan
Max Neal
Ken Shuttleworth

Consultants

Engineering Concept: Michel Virlogeux
Structural Engineer: Europe Etudes Gecti; Sogelerg; SERF
Landscape Architect: Agence TER
Associate Architects: Chapelet-Defol-Mousseigne
Contractor: Compagnie Eiffage du Viaduc de Millau (CEVM)

Principal Awards

2002 Singapore Construction Excellence Award – Civil category
2005 D&AD Gold Award
2005 Festival Automobile International Award
2005 RIBA Award
2005 Travel + Leisure Design Award for Best Infrastructure
2005 ECCS European Award for Steel Structures
2006 The Building Exchange (BEX) Award, 2nd Place in the Regeneration category
2006 Balthasar Neumann Prize – Commendation
2006 Staalbouwwedstrijd/Concours Construction Acier – Winner International category
2006 IABSE Outstanding Structure Award
2006 Wallpaper Design Awards Best New Public Building
2008 Chicago Athenaeum International Architecture Award

Project chronology

1987 Preliminary feasibility studies are undertaken for a bridge carrying the A75 motorway across the Tarn Gorge; four routing options are examined in detail: respectively, red, yellow, black and blue

1989 28 June: the blue route is selected, though two strategies remain open for further study: a high-level bridge across the Tarn Gorge, or a low-level bridge combined with a tunnel at its southern end on the Causse du Larzac

1991 October: the high-level route across the Tarn Gorge is officially selected
The structures division of Sétra, directed by Michel Virlogeux, examines the feasibility of a single structure spanning the valley; these studies are concluded in 1993

1993 July: an architectural/engineering competition is launched to design a bridge on the high-level route; 17 structural engineers and 38 architects indicate interest in the project; eight structural engineers are selected for the technical study and seven architects for the architectural study

1995 February: five bridge designs are shortlisted

1996 15 July: Minister of Public Works Bernard Pons announces that the multi-span cable-stayed solution presented by Europe Etudes Gecti, Sogelerg, SERF and Foster + Partners is the winning design

1998 May: it is decided that a concession for the bridge's construction and future operation will be tendered
Wind tunnel tests are undertaken and the detailed design of the bridge is finalised

2000 June: an international invitation to tender for the concession is issued; bids are received from four consortia

2001 August: Compagnie Eiffage du Viaduc de Millau (CEVM), a subsidiary of the French construction company Eiffage, is awarded the bridge concession
10 October: the decree awarding the bridge concession to Eiffage is published in the *Journal Officiel*
14 December: the first stone is laid by Minister of Transport Jean-Claude Gayssot

2002 10 January: construction of the southern abutment, C8, begins
February: work commences on the pier foundations
14 April: construction of the first pier, P6, begins
June: abutment C8 is complete
July: work begins on foundations for the temporary road deck support towers
September: assembly of the box girders that form the road deck begins
November: the first piers are complete

2003 21 February: P2 reaches 145 metres, breaking the French height record
25-26 March: the launching operation for the first southern sections of roadway commences
12 June: P2 reaches 183 metres, breaking the record as the world's highest pier
10 September: the launching operation for the first northerly sections of roadway begins
20 October: P2 is complete at 244.96 metres
24 November: the last of the seven piers, P3 (height 221.05 metres), is complete
9 December: a firework display celebrates the end of the nine concrete worksites

2004 7 April: HRH The Duke of Edinburgh visits the site; the two halves of the roadway are just metres apart and the southern section will be moved into place within three weeks
24 April: the sixth and final launch operation of the northern deck is completed
June: installation of the masts and supporting cable stays commences, followed by the removal of the temporary road deck supports
21-24 September: asphalt is laid on the road deck
6-25 November: the bridge is subjected to deadweight and dynamic testing to verify its strength
14 December: Monsieur Jacques Chirac, President of the French Republic, formally inaugurates the bridge
16 December: Millau Viaduct opens to traffic, one month ahead of schedule

Vital statistics

The bridge's construction broke several world records:

It has the highest piers in the world: piers P2 and P3, at 244.96 metres/ 803.5 feet and 221.05 metres/725 feet respectively, breaking the French record previously held by the Tulle and Verrières Viaducts (141 metres/462.5 feet), and the world record previously held by the Kochertal Viaduct in Germany, which is 181 metres/594 feet at its highest point

It has the highest bridge tower in the world: the mast atop pier P2 peaks at 343 metres/1,125 feet, replacing the Eiffel Tower (324 metres/1,063 feet) as France's tallest structure

It has the highest road bridge deck in Europe: at 270 metres/885.5 feet above the Tarn River at its highest point; it is nearly twice as tall as the previous tallest vehicular bridges in Europe, the Europabrucken in Austria and the Italia Viaduct in Italy

Its construction involved the world's longest ever bridge deck launch when it was 'pushed out' from the south abutment; this involved the use of the first synchronised set of computer-controlled horizontal jacks

Cost
: Approximately €320million, privately funded

Route
: The bridge spans from the Causse Rouge plateau to the north to the Causse Larzac plateau to the south, completing a missing link in the A75 autoroute – also known as la Méridienne – from Clermont-Ferrand to Béziers (and by extension Paris to Barcelona) across the Massif Central; the A75 now provides the shortest, fastest route by road from Paris to the Mediterranean coast

Total viaduct length
: 2.46km/1.53 miles

Spans between piers
: The six central spans measure 342 metres/1,122 feet; the two outer spans are 204 metres/ 669 feet

Height of road deck above River Tarn
: 270 metres/885.5 feet

Heights of the piers
: Pier 1: 94.50 metres/309.5 feet
: Pier 2: 244.96 metres/803.5 feet
: Pier 3: 221.05 metres/725 feet
: Pier 4: 144.21 metres/473 feet
: Pier 5: 136.42 metres/447.5 feet
: Pier 6: 111.94 metres/367 feet
: Pier 7: 77.56 metres/254.5 feet

Height to top of tallest mast
: 343 metres/1,125 feet

Base dimensions of tallest pier
: 25 x 17 metres/82 x 56 feet

Total construction weight
: 242,000 tonnes (36,000 tonnes of steel, 206,000 tonnes of concrete)

Deck launch
: A technically advanced system was designed to move the steel box-girder deck, as two continuous sections, across seven piers and seven temporary towers, each equipped with computer-controlled hydraulic jacks

: The two sections were launched from the north and south abutments respectively, the deck advancing a maximum 600mm every four minutes. In total, 3,280 push cycles from the west and 1,540 from the east were required to complete the entire span

Road deck
: The six central spans are composed of 173 central box beams, which form a spinal column on to which the lateral floors and the lateral box beams are welded

: The central box beams have a 4-metre/13-foot cross-section and a length of 15-22 metres/49-72 feet

The road deck slopes down at an angle of 3.025 per cent from south to north

The deck curves in plan, on a 20km/12.43 mile radius, to allow drivers to appreciate the overall structural form as they approach and pass over it

The deck has a trapezoidal section, which provides negative lift in strong wind conditions

Transparent aerodynamic side screens shelter vehicles from high gusts of wind

Masts

Seven steel masts, set on top of the piers, are 87 metres/285 feet high and each weighs approximately 700 tonnes

Between each mast, eleven cable stays (comprising 55 to 91 high-tensile-steel strands, depending on their length) fan down to support the deck; the stays are anchored in the central reservation between the carriageways

Traffic

On 22 July 2011 the 30-millionth vehicle crossed the viaduct – an average of 12,500 vehicles per day

Credits

Editor: David Jenkins
Design: Thomas Manss & Company; Thomas Manss, Tom Featherby, Angela Pescolderung
Picture Research: Gayle Mault, Lauren Catten
Proofreading: Julia Dawson, Rebecca Roke
Production Supervision: John Bodkin, Martin Lee
Reproduction: Dawkins Colour
Printed and bound in Italy by Grafiche SiZ S.p.A.

The FSC®-certified paper GardaMatt has been supplied by Cartiere del Garda S.p.A., Italy

FSC
www.fsc.org
MIX
Paper from responsible sources
FSC® C005613

Picture credits

Photographs
180° d'Images: 68
Jean-Philippe Arles/Reuters/Corbis: 78, 82, 83
Stephane Compoint: 25 (top), 50, 51, 55, 62, 63
Eiffage/Daniel Jamme: 17 (bottom), 29 (bottom), 58, 59, 60-61, 66-67, 79
Foster + Partners: 14 (r), 16 (top), 17 (top), 21, 24, 25 (bottom), 26-27
Getty Images: 14 (l)
Ben Johnson: 6-7, 69, 72-73, 74, 75, 76, 77, 80, 81
La Poste: 29 (top)
Colm Tamney: 86, 87
SBA73/Flickr: 31 (top)
Nigel Young/Foster + Partners: 16 (bottom), 18, 19, 20, 28, 30, 31 (bottom), 34-35, 36, 37, 38, 39, 42, 43, 44, 45, 46, 47, 48, 49, 54, 56-57, 88-89, 94-95

Drawings and Sketches
Foster + Partners: 22, 23, 70-71, 84-85
Norman Foster: 4-5
Gregory Gibbon/Foster + Partners: 12-13, 15, 32-33, 40-41, 52-53, 64-65

Editor's Note

In editing this book I am particularly grateful to Norman Foster and Thomas Leslie for their invaluable contributions. I would also like to thank Thomas Manss, Tom Featherby and Angela Pescolderung for bringing the book to life graphically; Gayle Mault and Lauren Catten, who mined the office archive; Julia Dawson and Rebecca Roke for proofreading the text; Martin Lee and John Bodkin for coordinating production; and the numerous people in the Foster studio – past and present – who helped piece together the background to the project.

David Jenkins
London, June 2012